The North Eastern Locomotives

in old picture postcards

A review of the North Eastern Railway Locomotive classes from A to Z, 1885-1923

by
J. Robin Lidster

European Library - Zaltbommel/Netherlands MCMLXXXV

Cover picture:
A handsome North Eastern tank engine (class W) simmers gently outside Scarborough shed in 1920. Most authors of books on North Eastern locomotives have concentrated on passenger locomotives and, to a greater or lesser extent, neglected the tank engines, particularly in their illustrations. The author hopes that this book will help to restore the balance.

About the author:
Robin Lidster was born in Scarborough in 1942 and was educated at Friends' School at Ackworth. He has worked at museums in Oxford, Huddersfield and Scarborough. His interest in the North Eastern Railway stems from his detailed researches into the history of the Scarborough & Whitby Railway, and its villages, and other railways on the Yorkshire coast.

Other books by Robin Lidster:
The Scarborough & Whitby Railway
Robin Hood's Bay As It Was
Yorkshire Coast Lines
Robin Hood's Bay in old picture postcards
From Scalby to Ravenscar in old picture postcards
The Scarborough & Whitby Railway – A Centenary Volume

For Molly & Charles Edward Moisley

GB   ISBN 90 288 3204 1 / CIP

European Library in Zaltbommel/Netherlands publishes among other things the following series:

IN OLD PICTURE POSTCARDS *is a series of books which sets out to show what a particular place looked like and what life was like in Victorian and Edwardian times. A book about virtually every town in the United Kingdom is to be published in this series. By the end of this year about 175 different volumes will have appeared. 1,250 books have already been published devoted to the Netherlands with the title* **In oude ansichten.** *In Germany, Austria and Switzerland 500, 60 and 15 books have been published as* **In alten Ansichten;** *in France by the name* **En cartes postales anciennes** *and in Belgium as* **En cartes postales anciennes** *and/or* **In oude prentkaarten** *150 respectively 400 volumes have been published.*

For further particulars about published or forthcoming books, apply to your bookseller or direct to the publisher.

This edition has been printed and bound by Grafisch Bedrijf De Steigerpoort in Zaltbommel/Netherlands.

INTRODUCTION

The North Eastern Railway Company was formed in 1854 by the amalgamation of the York, Newcastle & Berwick Railway, the York & North Midland Railway, the Leeds Northern Railway and some smaller companies. The new company eventually had the virtual monopoly of rail traffic in north-east England from Hull to Berwick.

The North Eastern acquired a very miscellaneous collection of locomotives from its constituent companies and for many years new engines were turned out, with little regard to standardisation, by the locomotive works at Shildon, Gateshead, Darlington and York. It was not until 1885, under the direction of the new Locomotive Superintendant, Thomas W. Worsdell, that serious efforts were made towards standardisation. Worsdell also simplified the external appearance of the locomotives which he provided with an impressive brass safety-valve casing which soon became the hallmark of North Eastern locomotives. The improved designs were perpetuated by his successors.

The locomotive works at Darlington and Gateshead were the only ones, belonging to the North Eastern, which constructed new engines during the period covered by this book. Gateshead Works was established in the 1840's and became the locomotive headquarters of the North Eastern in 1854. By 1910 there was no room for further expansion on the site and new headquarters were built at Darlington.

The Darlington Works were originally opened by the Stockton & Darlington Railway Company in 1863, the same year that the company was amalgamated with the North Eastern. There was plenty of space here and the works were expanded in 1884, 1903 and 1910. Many of the locomotive classes illustrated in this book were built at Darlington Works which closed in 1966.

The man ultimately responsible for the overall design of the locomotives, if not the details, was the Locomotive Superintendant, later known as Chief Mechanical Engineer.

The first of these, during the period covered by this book, was T.W. Worsdell who was born in 1838. He worked at Crewe where he became Works Manager in 1871 after a spell of five years with the Pennsylvania Railroad. He joined the Great Eastern Railway in 1881 where he experimented with compounding (see caption 2). In 1885 he became Locomotive Superintendant of the North Eastern Railway and produced about a dozen new classes which set the standard for North Eastern locomotive design for the next thirty years.

Wilson Worsdell succeeded his brother, T.W. Worsdell, in 1890. He was born at Crewe and went to Friends' School at Ackworth near Pontefract in Yorkshire. He too went to the Pennsylvania Railroad but joined the London & North Western Railway on his return to this country. During his time with the North Eastern he brought out about two dozen new locomotive designs.

Vincent L. Raven (later Sir) followed as Chief Mechanical Engineer, in 1910, but was distinguished more for his administrative abilities than his locomotive designs. His innovations on the North Eastern included locomotive cab signalling and the widespread use of superheating. Of his locomotives the class Z was said to be one of the best types of passenger locomotives in the country.

Another personality that stands out in North Eastern locomotive history is that of W.M. Smith who became Chief Draughtsman at Gateshead Works in 1883. Under Wilson Worsdell he was able to put his three-cylinder compound system into practice and it became the forerunner of a very successful class on the Midland Railway. They were not proliferated on the North Eastern because of the royalties demanded by Smith's executors.

This book covers the period from the end of 1885 to 1923, when the North Eastern became part of the London & North Eastern Railway. The first new locomotive type to appear under T.W. Worsdell was designated class A, in March 1886, and subsequent classes took the next available letter of the alphabet until class Z appeared in 1911.

The purpose of this book is to provide illustrations of all the lettered classes of locomotives, as well as some of the other contemporary motive power, during the period concerned. The author hopes that this will provide a useful and concise reference for postcard collectors, modellers and steam enthusiasts generally. It is not possible in a book of this size to provide more than a brief outline of each class but reference to works by the following authors will provide a wealth of technical and historical details: E.L. Ahrons, C.J. Allen, J.W. Armstrong, C.H. Ellis, K. Hoole, R.J. Irving, J.W. Lowe, J.S. McLean, O.S. Nock, W.A. Tuplin and J.N. Westwood. The main reference works are referred to by title in the captions.

Acknowledgements:

In addition to the above sources I would also like to thank the following for their help in providing illustrations and information: Derek Barker, Michael Brett, John L. Brown, Pete Cooper, Larry Dudding, Colin Foster, E.B. Longbottom, Irene Newham and Ken Taylor. My special thanks go to Willie Yeadon for his valuable assistance and encouragement. The helpful assistance of the staff of the National Railway Museum Library, at York, is also gratefully recorded. Most of the illustrations in this book are from old postcards collected by the late Charles Moisley, a Scarborough railway enthusiast. But for the kindness of his late widow in passing on his collection this book would not have been compiled.

Author's note:

The measurements and weights given for the locomotives in this book are only approximate as many of the dimensions and details varied from engine to engine, within a class, depending on wear and tear, modifications, rebuilding and many other factors. To take the apparently simple measurement of length, for example, this could vary according to the type of buffer beam and buffers and also on how closely the tender was coupled to its engine. The weights given by different authorities vary considerably and are sometimes the total weight in working order and sometimes the empty weight which can make a difference of a few tons. Despite the possibility of some inaccuracies creeping in I have given figures, where possible, in order that some comparison can be made between the different classes. Those who require accurate details should refer to the original drawings held by the various archives, to the excellent series of volumes 'Locomotives of the L.N.E.R.' by the Railway Correspondence & Travel Society, or to the few surviving locomotives of the former North Eastern Railway.

1. CLASS: A, NUMBER: 172, BUILT: 1888, SHED 1920: Tyne Dock, WITHDRAWN: 1929, CLASS INTRODUCED: 1886, NUMBER IN CLASS: 60.
LNER CLASS: F8, WHEEL ARRANGEMENT: 2-4-2, DRIVING WHEEL DIAMETER: 5½ Ft., LENGTH: 34¾ Ft., WEIGHT: 53¾ TONS.
T.W. Worsdell's first locomotive class for the North Eastern – a neat tank engine designed for short distance passenger traffic. The polished brass safety-valve casing became a standard feature on all new North Eastern engines for the next twenty-five years. The new boilers also became a standard design and were used on many of the smaller locomotives. The last engines were withdrawn in 1938.
NUMBERS: 13, 21, 35, 41, 55, 72, 128, 155, 169, 172, 187, 201, 205, 262, 279, 404, 414, 415, 418-20, 423, 425, 454, 469, 483, 485, 490, 507, 537, 575, 671, 674, 685, 801, 804, 854, 1160, 1171, 1322, 1577-86, 1597-1606.

2. CLASS: B, NUMBER: 219, BUILT: 1888, SHED 1920: Hull, WITHDRAWN: 1952, CLASS INTRODUCED: 1888, NUMBER IN CLASS: 51.

LNER CLASS: N8, WHEEL ARRANGEMENT: 0-6-2, DRIVING WHEEL DIAMETER: 5 Ft., LENGTH: 35 Ft., WEIGHT: 54¼ TONS.

One of the first of T.W. Worsdell's compound locomotive classes for the North Eastern. Under this system high pressure steam is fed to the smaller of two cylinders and, after moving the piston, instead of being exhausted, is used in the second, larger cylinder, at a lower pressure. This gave a more economical use of steam but it was difficult to balance the power output and often gave trouble in starting.

NUMBERS: 76, 136, 213, 215, 216, 218, 219, 238, 267, 271, 284, 287, 293, 345, 346, 348-51, 371, 373, 445, 503, 509, 515, 523, 573, 683, 780, 809, 855-64, 959, 961, 1072, 1091, 1104, 1105, 1124, 1127, 1145, 1152, 1168.

3. CLASS: B1, NUMBER: 428, BUILT: 1886, SHED 1920: East Hartlepool, WITHDRAWN: 1930, CLASS INTRODUCED: 1886, NUMBER IN CLASS: 11.
LNER CLASS: N8, WHEEL ARRANGEMENT: 0-6-2, DRIVING WHEEL DIAMETER: 5 Ft., LENGTH: 35 Ft., WEIGHT: 54¼ TONS.
B1, the simple version of this class, was actually produced in 1886 but they were identical in appearance to the class B engines. Both were designed for short-haul goods but were later used for passenger work. The class B were all rebuilt by Wilson Worsdell as simple engines by 1912 and all were known as class B, rather than B1, from 1914. The last engine was withdrawn in 1956. A useful reference is 'Locomotive Designers in the Age of Steam', by J.N. Westwood.
NUMBERS: (class B1) 14, 74, 185, 210, 212, 428, 504, 528, 531, 535, 1165.

4. CLASS: C, NUMBER: 104, BUILT: 1890, SHED 1920: Carlisle, WITHDRAWN: 1935, CLASS INTRODUCED: 1886, NUMBER IN CLASS: 171.

LNER CLASS: J21, WHEEL ARRANGEMENT: 0-6-0, DRIVING WHEEL DIAMETER: 5 Ft., LENGTH: 50 Ft., WEIGHT: 40½ TONS.

A tender version of the class B. These compound locomotives were originally designed for longer distance light freight.

NUMBERS: 16, 26, 30, 31, 34, 48, 51, 56, 68, 93, 95, 97, 99, 101, 102, 104, 107, 110, 122, 123, 133, 139, 147, 148, 157, 160, 209, 259, 289, 291, 294, 300, 312-16, 331, 424, 431, 432, 458, 470, 510, 511, 513, 520, 530, 534, 538, 556, 564, 568-70, 579, 582, 611, 613, 619, 665, 666, 668, 680, 776, 778, 800, 806, 807, 810, 869, 871, 872, 874-78, 899, 944, 960, 962, 963, 965, 973, 975, 976, 979, 981, 992-94, 996, 997, 1071, 1073, 1075, 1122, 1161, 1187, 1188, 1301, 1305, 1309, 1315, 1323, 1332, 1336-39, 1507-16, 1547-76, 1587-96, 1607-16.

5. CLASS: C1, NUMBER: 1808, BUILT: 1894, SHED 1920: York, WITHDRAWN: 1937, CLASS INTRODUCED: 1886, NUMBER IN CLASS: 30.

LNER CLASS: J21, WHEEL ARRANGEMENT: 0-6-0, DRIVING WHEEL DIAMETER: 5 Ft., LENGTH: 50 Ft., WEIGHT: 40½ TONS.

The simple version of the class C tender locomotive equivalent to the class B1 tank engine. These were the first tender locomotives designed by T.W. Worsdell for the North Eastern and O.S. Nock in his book 'Locomotives of the North Eastern Railway' records their favourable reception: *The spacious covered-in cab attracted much attention at the time, and it was suggested that the idea came from Mr. Worsdell's experience in America. While providing an excellent shelter the cab was handsomely finished inside, with the sidesheets and roof lined with wood, painted and polished.* Last engine withdrawn 1962.

NUMBERS: 22, 86, 152, 182, 360, 480, 539, 558, 667, 974, 1801-20.

6. CLASS: D, NUMBER: 340, BUILT: 1888, SHED 1920: Hull, WITHDRAWN: 1933, CLASS INTRODUCED: 1886, NUMBER IN CLASS: 2.
LNER CLASS: -, WHEEL ARRANGEMENT: 2-4-0, DRIVING WHEEL DIAMETER: 6¾ Ft., LENGTH: 53½ Ft., WEIGHT: 43¼ TONS.
T.W. Worsdell's first express passenger locomotives for the North Eastern. They were two-cylinder compounds on the Worsdell-von Borries system which he had tried out on the Great Eastern Railway. Both members of the class (1324 & 340) were rebuilt in 1896, as simple engines, when a leading bogie was fitted and they joined class F1. Number 340 was the first North Eastern locomotive to use piston valves, instead of the common slide valve, to feed the required amount of steam to the cylinders. Both engines were withdrawn in the early 1930's.

7. CLASS: D, NUMBER: 2143, BUILT: 1913, SHED 1920: Leeds, WITHDRAWN: 1960, CLASS INTRODUCED: 1913, NUMBER IN CLASS: 45.
LNER CLASS: H1, WHEEL ARRANGEMENT: 4-4-4, DRIVING WHEEL DIAMETER: 5¾ Ft., LENGTH: 43 Ft., WEIGHT: 84¾ TONS.
After the rebuilding of 1324 and 340 the D classification remained vacant until this three-cylinder tank engine was produced by Vincent Raven in 1913. The 4-4-4T wheel arrangement was unusual for the North Eastern and the engines were designed for passenger work on longer-distance branch lines. They were fitted with steam reversing gear – the apparatus can be seen just in front of the side-tank. Starting in 1931 the LNER rebuilt them as 4-6-2T and they became very popular as class A8. Many survived until 1960.
NUMBERS: 1326-30, 1499-1503, 1517-1531, 2143-2162.

8. CLASS: E, NUMBER: 811, BUILT: 1887, SHED 1920: Borough Gardens, WITHDRAWN: 1937, CLASS INTRODUCED: 1886, NUMBER IN CLASS: 120.

LNER CLASS: J71, WHEEL ARRANGEMENT: 0-6-0, DRIVING WHEEL DIAMETER: 4½ Ft., LENGTH: 27¾ Ft., WEIGHT: 37½ TONS.

A neat little tank engine designed for light freight and station shunting by T.W. Worsdell. Many survived into the 1950's.

NUMBERS: 27, 50, 54, 70, 77, 84, 103, 137, 144, 161, 165, 168, 176, 177, 179, 181, 221, 224, 225, 237, 239-42, 244, 248, 252, 254, 260, 261, 263, 268, 272, 275, 278, 280, 285, 286, 296, 299, 301, 304, 317, 326, 338, 347, 399-403, 447-53, 478, 482, 492-96, 499, 501, 533, 541, 572, 57, 584, 802, 811, 969, 972, 977, 978, 980, 1083-85, 1095, 1103, 1123, 1134, 1140, 1142-44, 1151, 1153, 1155, 1157, 1163, 1167, 1196-99, 1314, 1666, 1688-90, 1735, 1758, 1789, 1796, 1797, 1831-36, 1861-64.

9. CLASS: E1, NUMBER: 1720, BUILT: 1899, SHED 1920: York, WITHDRAWN: 1961, CLASS INTRODUCED: 1898, NUMBER IN CLASS: 75.
LNER CLASS: J72, WHEEL ARRANGEMENT: 0-6-0, DRIVING WHEEL DIAMETER: 4 Ft., LENGTH: 27¾ Ft., WEIGHT: 36¼ TONS.

The E1 class of Wilson Worsdell, introduced in 1898, was a development of T.W. Worsdell's class E. The boiler was identical but the cylinders were enlarged and the wheels reduced in diameter. This class became the standard North Eastern shunting type and further batches were built by the LNER (10) and British Railways (28). In 1921 the cost of a new engine of this class, built by Armstrong Whitworth & Co., was almost £7,000. The first engines were not withdrawn until 1958, the last in 1967. The cost of the first engine, 1720, was £1,412.
NUMBERS: 462, 1715, 1718, 1720-22, 1728, 1732-34, 1736, 1741, 1742, 1744, 1746, 1747, 1749, 1761, 1763, 1770, 2173-92, 2303-37.

10. CLASS: F, NUMBER: 18, BUILT: 1887, SHED 1920: Gateshead, WITHDRAWN: 1929, CLASS INTRODUCED: 1886, NUMBER IN CLASS: 25.
LNER CLASS: D22, WHEEL ARRANGEMENT: 4-4-0. DRIVING WHEEL DIAMETER: 6¾ Ft., LENGTH: 53½ Ft., WEIGHT: 46½ TONS.
A development of the original class D but with a leading bogie and compound cylinders. McLean records that engine 117 took part in the Races to Edinburgh in 1888 and made the record run from Newcastle, a distance of 124¼ miles, in 126 minutes. Other authors give three different timings for this run and one even suggests that 117 was piloting another locomotive at the time! The last engine was withdrawn from service in 1935. The photograph shows number 18 at Scarborough.
NUMBERS: 18, 42, 115, 117, 355, 356, 514, 663, 684, 779, 1532-46.

11. CLASS: F1, NUMBER: 1324, BUILT: 1886, SHED 1920: Gateshead, WITHDRAWN: 1930, CLASS INTRODUCED: 1887, NUMBER IN CLASS: 10+27.
LNER CLASS: D22, WHEEL ARRANGEMENT: 4-4-0, DRIVING WHEEL DIAMETER: 6¾ Ft., LENGTH: 53 Ft., WEIGHT: 46 TONS.
Class F1 was the non-compound version of class F with two high-pressure cylinders 18 inches in diameter. The boilers and motion were exactly the same so that an accurate comparison could be made between the compound and simple systems. Wilson Worsdell rebuilt the class F and original class D engines, 27 in all, to class F1 from 1896 to 1905. The ten compound locomotives were initially employed on the East Coast expresses but the class F1 worked secondary passenger trains. They were all scrapped by 1936. The photograph shows 1324 (originally class D, see caption 6) at Benton Quarry in 1919, Ken Nunn/LCGB Collection.
NUMBERS: (class F1) 85, 96, 154, 194, 230, 673, 777, 803, 808, 1137.

12. CLASS: 1463, NUMBER: 1469, BUILT: 1885, SHED 1920: York, WITHDRAWN: 1927, CLASS INTRODUCED: 1885, NUMBER IN CLASS: 20.
LNER CLASS: E5, WHEEL ARRANGEMENT: 2-4-0, DRIVING WHEEL DIAMETER: 7 Ft., LENGTH: 50¾ Ft., TOTAL WEIGHT: 42 TONS.
The 1463 class was introduced in 1885 during the short period after the resignation of the previous Locomotive Superintendant (A. McDonnell) and the appointment of T.W. Worsdell, with whom we started this book. Although designed under the auspices of a Locomotive Committee under the chairmanship of the General Manager, Henry Tennant, T.W. Worsdell had a lot to do with the design being one of the Assistant Locomotive Superintendants at the time. It is placed here, next to the original form of class G, for comparison. 1463 is preserved by the National Railway Museum.
NUMBERS: 1463-79, 1504-6.

13. CLASS: G, NUMBER: 676, BUILT: 1887, SHED 1920: Hull, WITHDRAWN: 1930, CLASS INTRODUCED: 1887, NUMBER IN CLASS: 20.
LNER CLASS: - , WHEEL ARRANGEMENT: 2-4-0, DRIVING WHEEL DIAMETER: 6 Ft., LENGTH: 50¼ Ft., WEIGHT: 40¼ TONS.
A small passenger engine, designed by T.W. Worsdell, for branch line work. Apart from the two members of the original class D these were the only 2-4-0's built by T.W. Worsdell. McLean records that they were nicknamed 'Waterbury's', after a well-known make of watch, because they kept good time on passenger trains. Other authors say the footplate staff were far from enthusiastic about them. So many factors could affect the performance of a locomotive and not least the effects of scale deposit on the inside of the boiler (like the 'fur' in a kettle)...

14. CLASS: G1, NUMBER: 223, BUILT: 1888, SHED 1920: Hull, WITHDRAWN: 1933, CLASS INTRODUCED: 1887, NUMBER IN CLASS: 20.
LNER CLASS: D23, WHEEL ARRANGEMENT: 4-4-0, DRIVING WHEEL DIAMETER: 6 Ft., LENGTH: 52 Ft., WEIGHT: 44¼ TONS.
...The North Eastern Railway Magazine of 1912 (pp 239-241) contains an interesting article which gives the increase in fuel consumption in proportion to the thickness of scale − 15% at 1/16th of an inch up to 60% at ¼ inch. The engines were all rebuilt by Wilson Worsdell from 1900 to 1904. The cylinders were enlarged and a 4-wheel bogie replaced the leading pair of wheels. They were further modified and superheated by Vincent Raven and the last of the class was scrapped in 1935.
NUMBERS: 23, 214, 217, 222, 223, 258, 274, 328, 337, 372, 472, 521, 557, 675-9, 1107, 1120.

15. CLASS: H, NUMBER: 24, BUILT: 1888, SHED 1920: Percy Main, WITHDRAWN: 1931, CLASS INTRODUCED: 1888, NUMBER IN CLASS: 19.
LNER CLASS: Y7, WHEEL ARRANGEMENT: 0-4-0, DRIVING WHEEL DIAMETER: 3½ Ft., LENGTH: 20 Ft., WEIGHT: 21¾ TONS.
Small shunting tank engines which were used on the North Eastern docks. Apart from class K these had the smallest boilers of any North Eastern locomotives built during this period with a total heating surface of only 505 square feet compared with class V with 2,456 square feet. Five more of the class were built by the LNER in 1923 (Numbers 982-986). The last engine was withdrawn in 1952 but some went into private ownership and lasted many years longer. Photograph: A.G. Ellis Collection.
NUMBERS: 24, 129, 518, 519, 587, 898, 900, 945, 946, 1302-4, 1306-8, 1310, 1798-1800.

16. CLASS: H1, NUMBER: 995, BUILT: 1888, SHED 1920: Gateshead, WITHDRAWN: 1933, CLASS INTRODUCED: 1888, NUMBER IN CLASS: 2.
LNER CLASS: J78, WHEEL ARRANGEMENT: 0-6-0, DRIVING WHEEL DIAMETER: 3½ Ft., LENGTH: 23¼ Ft., WEIGHT: 26½ TONS.
The two locomotives of this class were specifically built for use at Gateshead Works although 590 was stationed at York for some years. 995 had cylinders 14 inches in diameter and 590 13 inches. This was an extended version of class H with an extra pair of wheels which helped to carry the weight of a steam-powered crane with a lifting capacity of 3 tons. The reason for the lack of a dome on these boilers is clear from the photograph. Both locomotives were withdrawn in the 1930's. Details of some of the smaller classes are given in 'North Eastern Steam' by W.A. Tuplin.
NUMBERS: 590, 995.

17. CLASS: H2, NUMBER: 1787, BUILT: 1897, SHED 1920: Middlesbrough, WITHDRAWN: 1936, CLASS INTRODUCED: 1897, NUMBER IN CLASS: 3.
LNER CLASS: J79, WHEEL ARRANGEMENT: 0-6-0, DRIVING WHEEL DIAMETER: 3½ Ft., LENGTH 22 Ft., WEIGHT: 25 TONS.
The craneless version of class H1 which again utilised the domeless boiler common to the other two classes. These locomotives were used for shunting but did work short distance passenger trains for a while. After withdrawal 1787 went to the Bowes Railway, in 1936, and was scrapped in 1946. 407 went to a colliery and 1662 went to a sugar beet factory at Norwich. In 1920 the engines were stationed at Darlington, Gateshead and Middlesbrough. The book 'British Steam Locomotive Builders', by James W. Lowe, is a mine of information on North Eastern and other locomotives.
NUMBERS: 407, 1662, 1787.

18. CLASS: I, NUMBER 1329, BUILT: 1888, SHED 1920: Hull, WITHDRAWN: 1921, CLASS INTRODUCED: 1888, NUMBER IN CLASS: 10.
LNER CLASS: w/d, WHEEL ARRANGEMENT: 4-2-2, DRIVING WHEEL DIAMETER: 7 Ft., LENGTH: 52 Ft., WEIGHT: 43¾ TONS.
Single driving-wheel compound express locomotives of a type which was unusual for the period. They were designed by T.W. Worsdell for fairly light trains on secondary lines and worked between Scarborough-Leeds and Hull-York for a number of years. These were the first 'singles' built by the North Eastern in 25 years and followed the success of a similar type on the Midland Railway in conjunction with the invention of sand blast apparatus to increase the friction on the rails. This was very necessary where the entire power of the locomotive was transmitted through a single pair of wheels.
NUMBERS: 1326-30, 1527-31.

19. CLASS: I, NUMBER 1528, BUILT: 1890, SHED 1920: w/d, WITHDRAWN: 1920, CLASS INTRODUCED: 1888, NUMBER IN CLASS: 10.
LNER CLASS: w/d, WHEEL ARRANGEMENT: 4-2-2, DRIVING WHEEL DIAMETER: 7 Ft., LENGTH 52 Ft., WEIGHT: 43¾ TONS.
As with most of T.W. Worsdell's compound designs Wilson Worsdell converted the ten class I engines to simples from 1900 to 1903. At the same time outside frames and bearings were added to the trailing wheels, beneath the firebox, which helped to remove a certain amount of unsteadiness inherent in the single driving-wheel arrangement. An excellent article, 'The Modern Locomotives of the North Eastern Railway', was published in the Railway Magazine of 1910 (pp 487-496). This is well-illustrated and covers the period 1885-1910. The photograph was taken at Scarborough.

20. CLASS: J, NUMBER: 1517, BUILT: 1889, SHED 1920: w/d, WITHDRAWN: 1920, CLASS INTRODUCED: 1889, NUMBER IN CLASS: 10.
LNER CLASS: w/d, WHEEL ARRANGEMENT: 4-2-2, DRIVING WHEEL DIAMETER: 7½ Ft., LENGTH: 53¾ Ft., WEIGHT: 46¾ TONS.
A larger and more powerful version of the class I single-wheeler compound with massive driving wheels 7 Ft. 7 inches in diameter. J.S. McLean records: *Engine 1517, the pioneer of this new class, began regular running on October 5th 1889, taking trains of from 10 - 22 coaches unaided. On Mr. Worsdell's own authority this class was built for running the heavy main line trains between Newcastle and Edinburgh. As compounds some really wonderful performances were achieved by class J, and engine 1518 was authentically timed at 86 m.p.h. — the first instance of so high a speed being recorded in Britain.*

21. CLASS: J, NUMBER: 1522, BUILT: 1890, SHED 1920: Hull, WITHDRAWN: 1921, CLASS INTRODUCED: 1889, NUMBER IN CLASS: 10,
LNER CLASS: w/d, WHEEL ARRANGEMENT: 4-2-2, DRIVING WHEEL DIAMETER: 7½ Ft., LENGTH: 53¾ Ft., WEIGHT: 46¾ TONS.
The low pressure cylinder on the compound class J was the largest used by the North Eastern at 28 inches in diameter. The low and high pressure cylinders had to be at different heights between the frames and the steam chests were outside the engine frame. This led to problems and they were all early candidates for conversion to the simple system in 1895/96. Outside bearings and frames were later added to the trailing wheels and they continued to perform well on trains between Scarborough and Leeds. They were all withdrawn from service between 1918 and 1920.
NUMBERS: 1517-1526.

22. CLASS: K, NUMBER n/k, BUILT: 1890, SHED 1920: Hull, WITHDRAWN: - , CLASS INTRODUCED: 1890, NUMBER IN CLASS: 5.
LNER CLASS: Y8, WHEEL ARRANGEMENT: 0-4-0, DRIVING WHEEL DIAMETER: 3 Ft., LENGTH: 17½ Ft., WEIGHT: 15½ TONS.
The last locomotive class to appear under the Superintendancy of T.W. Worsdell. These tiny engines were designed for use on Hull Docks and originally had a type of marine boiler. From 1902 -04 locomotive boilers, similar to class H, were used. A note in the North Eastern Railway Magazine (1914, p 39) describes the largest and smallest shunting engines in use at Hull (class X and class K) and is entitled 'Dignity and Impudence'. The last survivor of the class was 560 which was York shed pilot until 1956. The Library Association book 'Steam Locomotive Development', by Kevin P. Jones, provides a useful bibliography to the literature. NUMBERS: 559-563.

23. CLASS: L, NUMBER 553, BUILT: 1892, SHED 1920: Ferryhill, WITHDRAWN: 1960, CLASS INTRODUCED: 1891, NUMBER IN CLASS: 10.
LNER CLASS J73, WHEEL ARRANGEMENT: 0-6-0, DRIVING WHEEL DIAMETER: 4½ Ft., LENGTH: 31½ Ft., WEIGHT: 46¾ TONS.
This was the first new locomotive class to appear under Wilson Worsdell. The engines were designed as bankers to assist freight trains on steep gradients. The interesting book by R.J. Irving – 'The N E R Company 1870-1914 An Economic History' – gives some useful statistics and figures relating to the building and operation of North Eastern engines. In 1891, the year the class L was introduced, 82 new locomotives were built and the wages for the men involved in their construction worked out at £11 18s 9d. per ton. Details about the engine sheds can be found in 'North Eastern Locomotive Sheds' by K. Hoole.
NUMBERS: 544-553.

24. CLASS: M1, NUMBER: 1623, BUILT: 1893, SHED 1920: Carlisle, WITHDRAWN: 1932, CLASS INTRODUCED: 1892, NUMBER IN CLASS: 20.
LNER CLASS: D17/1, WHEEL ARRANGEMENT: 4-4-0, DRIVING WHEEL DIAMETER: 7 Ft., LENGTH: 55½ Ft., WEIGHT: 50¾ TONS.
Wilson Worsdell used large coupled wheels on this express passenger engine class after the good performance of the I and J classes. Twenty were built as simple engines plus one compound (class M − see caption 37) for comparative purposes. All the class M1 were designed to be easily rebuilt as compound engines but this did not take place as they proved to be excellent engines. These were the heaviest tender locomotives built so far during this period, weighing a total of over 90 tons with the tender. The last engine was withdrawn in 1945.
NUMBERS: 1620-1639.

25. CLASS: M1, NUMBER: 1630, BUILT: 1893, SHED 1920: Hull, WITHDRAWN: 1933, CLASS INTRODUCED: 1892, NUMBER IN CLASS: 20.
LNER CLASS: D17/1, WHEEL ARRANGEMENT: 4-4-0, DRIVING WHEEL DIAMETER: 7 Ft., LENGTH: 55½ Ft., WEIGHT: 50¾ TONS.

J.S. McLean records that these engines *bore the brunt of the North Eastern's share in the race to Aberdeen in 1895. On the night of August 21st the East Coast Companies made a special effort, and although the train consisted of six heavy six-wheeled joint stock coaches, engine 1621 ran from York to Newcastle in 78½ minutes, an average speed of 61½ mph – a truly fine performance which was immediately eclipsed by a sister engine, No. 1620, which created a world's record by running from Newcastle to Edinburgh at an average speed of 66 mph.* Engine 1621 is preserved in the National Railway Museum at York.

26. CLASS: N, NUMBER: 1651, BUILT: 1893, SHED 1920: West Auckland, WITHDRAWN: 1951, CLASS INTRODUCED: 1893, NUMBER IN CLASS: 20.

LNER CLASS: N9, WHEEL ARRANGEMENT: 0-6-2, DRIVING WHEEL DIAMETER: 5 Ft., LENGTH: 36 Ft., WEIGHT: 56½ TONS.

These engines were similar to class B1 but with a larger piston diameter and longer stroke although the same type of boiler was used. There is a very interesting article in the North Eastern Railway Magazine of 1912 (pp 161-3) entitled 'The Growth of Locomotive Boilers'. This illustrates many of the North Eastern boiler types as used on the various classes dealt with in this book. These range from the tiny class K boiler only 9 ft. long to that used on classes S1, V and Z which was 25 ft. long. An account of North Eastern engine construction can be found in the book by K. Hoole – 'North Road Locomotive Works Darlington 1863-1966'. Photograph – Ken Nunn/LCGB collection.

27. CLASS: N, NUMBER: 1655, BUILT: 1893, SHED 1920: Hartlepool, WITHDRAWN: 1955, CLASS INTRODUCED: 1893, NUMBER IN CLASS: 20.
LNER CLASS: N9, WHEEL ARRANGEMENT: 0-6-2, DRIVING WHEEL DIAMETER: 5 Ft., LENGTH: 36 Ft., WEIGHT: 56½ TONS.
The class N engines were designed for short distance goods train work but some were later adapted to work passenger trains by providing them with continuous brakes. J.W. Armstrong, in his book 'An Enthusiast Looks Back', records that *Class N was used on local passenger work but never seemed to run as sprightly as the earlier B and B1 engines. They were also employed at West Auckland and Barnard Castle for 'shoving up' to Stainmore Summit.* The last engine was withdrawn in 1955.
NUMBERS: 383, 1617, 1618, 1640-1655, 1705.

28. CLASS: O, NUMBER: 1701, BUILT: 1894, SHED 1920: Scarborough, WITHDRAWN: 1955, CLASS INTRODUCED: 1894, NUMBER IN CLASS: 110.
LNER CLASS: G5, WHEEL ARRANGEMENT: 0-4-4, DRIVING WHEEL DIAMETER: 5 Ft., LENGTH: 35 Ft., WEIGHT: 53¼ TONS.
The class O was a very successful design of tank engine which was used on a great variety of work on the North Eastern system. Some were to be found shunting in yards, others working local passenger traffic in the Leeds and Newcastle areas, and some 'mountaineering' in the Whitby district including the 46 mile coast route from Scarborough to Saltburn. They were pushed to the limit as J.W. Armstrong records: *One of the Starbeck engines did a tour round the West Riding (of Yorkshire) that used the engine to full capacity and... on occasion...*

29. CLASS: O, NUMBER: 1779, BUILT: 1896, SHED 1920: Middlesbrough, WITHDRAWN: 1954, CLASS INTRODUCED: 1894, NUMBER IN CLASS: 110.
LNER CLASS: G5, WHEEL ARRANGEMENT: 0-4-4, DRIVING WHEEL DIAMETER: 5 Ft., LENGTH: 35 Ft., WEIGHT: 53¼ TONS.
...the fireman would have to sweep the bunker out for coal! The engines were very smart in appearance and, although completely different in design, they had the same cylinders and boilers as the class A locomotives.
NUMBERS: 149, 380, 381, 384, 387, 394, 405; 408, 413, 427, 433, 435-37, 439, 441, 468, 505, 526, 529, 540, 580, 1096, 1169, 1316, 1319, 1334, 1687, 1691-93, 1695, 1701-3, 1713, 1730, 1737, 1739, 1740, 1745, 1748, 1751, 1752, 1754, 1755, 1759, 1762, 1764, 1765, 1769, 1772, 1775, 1778-80, 1783, 1786, 1788, 1791, 1793, 1795, 1837-40, 1865-68, 1881-90, 1911-20, 2081-2100.

30. CLASS: P, NUMBER 1897, BUILT: 1896, SHED 1920: Middlesbrough, WITHDRAWN: 1951, CLASS INTRODUCED: 1894, NUMBER IN CLASS: 70.
LNER CLASS: J24, WHEEL ARRANGEMENT: 0-6-0, DRIVING WHEEL DIAMETER: 4½ Ft., LENGTH: 49¼ Ft., WEIGHT: 38½ TONS.
A large proportion of the traffic on the North Eastern was in minerals and freight engines outnumbered passenger engines by three to one. The four varieties of class P were introduced by Wilson Worsdell to take over from the various ageing mineral engines of earlier decades. The class P locomotives also used the same boiler as the class A tank engines and the design owes part of its origin to the standard class C1. The last engine was withdrawn in 1951. The classic work on locomotives is undoubtedly 'The British Steam Railway Locomotive 1825-1925' by E.L. Ahrons.
NUMBERS: 1821-30, 1841-60, 1891-1900, 1931-60.

31. CLASS: P1, NUMBER: 2060, BUILT: 1900, SHED 1920: Whitby, WITHDRAWN: 1944, CLASS INTRODUCED: 1898, NUMBER IN CLASS: 120.

LNER CLASS: J25, WHEEL ARRANGEMENT: 0-6-0, DRIVING WHEEL DIAMETER: 4½ Ft., LENGTH: 51 Ft., WEIGHT: 41¾ TONS.

The first development of the class P was very modest – the cylinders were slightly larger and the boiler a few inches longer. It should be noted that the use of a number after the class letter is to some extent misleading, as, generally speaking, in the earlier years covered by this book it was purely intended to distinguish between compound and simple versions of the same class. In later years, as here, it was used to denote different stages in the development of an original class. After all the earlier compound engines had been converted to simples the suffix was often dropped.

NUMBERS: 25, 29, 257, 459, 463, 536, 1714, 1723-27, 1743, 1961-2000, 2031-80, 2126-42.

32. CLASS: P2, NUMBER: 1678, BUILT: 1904, SHED 1920: Newport, WITHDRAWN: 1959, CLASS INTRODUCED: 1904, NUMBER IN CLASS: 50.
LNER CLASS: J26, WHEEL ARRANGEMENT: 0-6-0, DRIVING WHEEL DIAMETER: 4½ Ft., LENGTH: 52¼ Ft., WEIGHT: 46¾ TONS.
The second development of the class P was on a more massive scale with a boiler of 5½ Ft. in diameter, larger than anything that had appeared previously on the North Eastern. The original working pressure was 200 lb. per square inch and the boiler carried four safety valves which were surrounded by a huge brass casing. First engine withdrawn in 1958.
NUMBERS: 67, 132, 233, 243, 342, 379, 406, 412, 434, 438, 442, 517, 525, 543, 554, 555, 765, 816, 818, 831, 835, 881, 1043, 1057, 1098, 1130, 1131, 1139, 1146, 1159, 1172, 1194, 1200, 1202, 1208, 1360, 1366, 1369, 1370, 1390, 1670, 1671, 1673, 1674, 1676, 1678, 1698, 1773, 1777, 1781.

33. CLASS: P3, NUMBER: 1212, BUILT: 1908, SHED 1920: Selby, WITHDRAWN: 1965, CLASS INTRODUCED: 1906, NUMBER IN CLASS: 115.
LNER CLASS: J27, WHEEL ARRANGEMENT: 0-6-0, DRIVING WHEEL DIAMETER: 4½ Ft., LENGTH: 52 Ft., WEIGHT: 47¾ TONS.
The final development of the class P, the P3, was very similar to the P2 class. Later engines of this class were constructed to an altered design of Sir Vincent Raven and had superheaters and Ross 'pop' safety valves. The last batch of 10 was not completed until 1923 after the North Eastern had been incorporated into the new London & North Eastern Railway Company.
NUMBERS: 790, 814, 836, 839, 880, 883, 888, 891, 917, 938, 1001, 1003-8, 1010-18, 1022-25, 1027-30, 1034-36, 1039, 1040, 1044, 1046-50, 1052, 1053, 1056, 1060, 1061, 1064-67, 1189, 1201, 1203-5, 1211-14, 1216, 1219-22, 1224-31, 1256, 1393, 1402, 1686, 2338-62, 2383-92.

34. CLASS: Q, NUMBER: 1902, BUILT: 1897, SHED 1920: Scarborough, WITHDRAWN: 1948, CLASS INTRODUCED: 1896, NUMBER IN CLASS: 30.
LNER CLASS: D17/2, WHEEL ARRANGEMENT: 4-4-0, DRIVING WHEEL DIAMETER: 7 Ft., LENGTH: 55½ Ft., WEIGHT: 49½ TONS.

A unique design, on the North Eastern, with a clerestory roof on the cab. These were intended for express passenger work but had smaller boilers than class M. C.H. Ellis, in 'Some Classic Locomotives', records that they performed well: *One of the class, engine number 1904, holds what I believe must be the record between York and Scarborough viz. the run with a special express of 137 tons with which the 42 miles were covered in 41 minutes.* This was during the trials for special seaside expresses and is better than present day timings.

NUMBERS: 1871-80, 1901-10, 1921-30.

35. CLASS: Q, NUMBER: 1901, BUILT: 1897, SHED 1920: Carlisle, WITHDRAWN: 1945, CLASS INTRODUCED: 1896, NUMBER IN CLASS: 30.
LNER CLASS: D17/2, WHEEL ARRANGEMENT: 4-4-0, DRIVING WHEEL DIAMETER: 7 Ft., LENGTH: 55½ Ft., WEIGHT: 49½ TONS.

The photograph shows two class Q locomotives in the North Eastern bay at Carlisle Citadel station. O.S. Nock says of this class: *For beauty of appearance the Q and Q1 engines must stand second to none among nineteenth century designs.* Part of the attraction of these, and other North Eastern locomotives, was undoubtedly their splendid livery of pale green with black and white border lining. This style was introduced by Wilson Worsdell in 1894 and went well with the polished brass safety-valve casing. The last engine was withdrawn in 1946.

36. CLASS: Q1, NUMBER: 1870, BUILT: 1896, SHED 1920: Leeds, WITHDRAWN: 1930, CLASS INTRODUCED: 1896, NUMBER IN CLASS: 2.
LNER CLASS: D18, WHEEL ARRANGEMENT: 4-4-0, DRIVING WHEEL DIAMETER: 7½ Ft., LENGTH: 56 Ft., WEIGHT: 50¾ TONS.
These impressive locomotives were designed by Wilson Worsdell specifically for racing. At the time they were said to have the largest coupled wheels in the world at 7 ft. 7 inches in diameter. After the rivalry between the East and West Coast routes for the Anglo-Scottish traffic in 1895 it was anticipated that there would be further competition in 1896. This did not develop but the locomotives continued to work the Scotch Expresses until 1906 and were capable of speeds of 70 mph. Both engines were withdrawn in 1930.
NUMBERS: 1869 and 1870.

37. CLASS: 3CC, NUMBER: 1619, REBUILT: 1898, SHED 1920: Hull, WITHDRAWN: 1930, CLASS INTRODUCED: 1898, NUMBER IN CLASS: 1.
LNER CLASS: D19, WHEEL ARRANGEMENT: 4-4-0, DRIVING WHEEL DIAMETER: 7 Ft., LENGTH: 56 Ft., WEIGHT: 53 TONS.
Originally this locomotive was built as class M, the two cylinder compound version of class M1 (see captions 24 and 25), in 1893 with inside cylinders. In 1898 it was rebuilt as a three-cylinder balanced compound to the designs of W.M. Smith, the Chief Draughtsman of the North Eastern. The high pressure cylinder was between the frames with the two low pressure cylinders on the outside. When starting a heavy load or working up an incline a valve operated by the driver allowed steam at full pressure into all three cylinders. The engine was quite a successful experiment and ran 21,000 miles in the first three months of its life.

38. CLASS: 290, NUMBER: 1346, BUILT: 1875, SHED 1920: York, WITHDRAWN: 1958, REBUILT: 1902, NUMBER IN CLASS: 60.
LNER CLASS: J77, WHEEL ARRANGEMENT: 0-6-0, DRIVING WHEEL DIAMETER: 4 Ft., LENGTH: 29¼ Ft., WEIGHT: 41¾ TONS.

Although this book deals almost exclusively with new engines built between 1885 and 1923 two or three of the more significant rebuilds are included. Number 1346 is one of these being originally a member of Edward Fletcher's BTP 0-4-4 tank engine class. Many of these were displaced when the class O tank engines were introduced and so a number were rebuilt as 0-6-0T for shunting work. York rebuilt forty of the engines between 1899 and 1904 which kept the rounded cab roof of Fletcher's time. Later rebuilds were given the typical square-cornered cab of the Worsdell type as shown in the photograph opposite...

39. CLASS: 290, NUMBER: 954, BUILT: 1874, SHED 1920: West Auckland, WITHDRAWN: 1960, REBUILT: 1921, NUMBER IN CLASS: 60.

LNER CLASS: J77, WHEEL ARRANGEMENT: 0-6-0, DRIVING WHEEL DIAMETER: 4 Ft., LENGTH: 29¾ Ft., WEIGHT: 43 TONS.

...Although classed as shunting locomotives some of them had to put in some hard work. O.S. Nock records that some worked on staiths at Blyth in the 1950's pushing loads of about 500 tons. Illustration 62 shows one of the BTP class in its original 0-4-4T form. Many of the rebuilt engines survived into the 1950's and some into the final years of steam on British Railways.

NUMBERS: 15, 37, 43, 47, 57, 71, 105, 138, 145, 151, 164, 166, 173, 199, 276, 290, 305, 319, 324, 333, 344, 354, 597, 604, 607, 612, 614, 623, 948, 953, 954, 956, 958, 998-1000, 1021, 1033, 1115, 1116, 1313, 1340-44, 1346-49, 1430-33, 1435, 1438, 1439, 1460-62.

40. CLASS: R, NUMBER: 2101, BUILT 1900, SHED 1920: Darlington, WITHDRAWN: 1956, CLASS INTRODUCED: 1899, NUMBER IN CLASS: 60.
LNER CLASS: D20, WHEEL ARRANGEMENT: 4-4-0, DRIVING WHEEL DIAMETER: 7 Ft., LENGTH: 56½ Ft., WEIGHT: 52 TONS.
This new design, with a larger boiler than any other class on the North Eastern at the time, proved very successful. The engines worked express passenger trains on the East Coast main line for many years. Engine 2011, of this class, was manned by two sets of footplate staff and in the first two years of its life it was said to have run nearly 140,000 miles. The total weight of the engine and tender, in working order, was just over 100 tons. These were the first North Eastern locomotives to be built with piston valves although many earlier engines had been rebuilt with this type.

41. CLASS: R, NUMBER 476, BUILT: 1906, SHED 1920: Darlington, WITHDRAWN: 1951, CLASS INTRODUCED: 1899, NUMBER IN CLASS: 60.
LNER CLASS: D20, WHEEL ARRANGEMENT: 4-4-0, DRIVING WHEEL DIAMETER: 7 Ft., LENGTH: 56½ Ft., WEIGHT: 54 TONS.
These were said to be the most famous engines introduced by Wilson Worsdell. Boiler pressure was 200 lb per square inch and the overall design was simpler and more reliable than many of the previous types of express passenger locomotives. The continuous splasher over the driving wheels had brass beading giving the appearance of separate splashers. Number 476, one of the last batch built, was virtually to the same design as the earlier batches. The last engine was withdrawn in 1957.
NUMBERS: 476, 592, 707, 708, 711-13, 723-25, 1026, 1042, 1051, 1078, 1147, 1184, 1206, 1207, 1209, 1210, 1217, 1223, 1232, 1234-36, 1258, 1260, 1665, 1672, 2011-30, 2101-10.

42. CLASS: R1, NUMBER: 1238, BUILT: 1908, SHED 1920: York, WITHDRAWN: 1945, CLASS INTRODUCED: 1908, NUMBER IN CLASS: 10.
LNER CLASS: D21, WHEEL ARRANGEMENT: 4-4-0, DRIVING WHEEL DIAMETER: 7 Ft., LENGTH: 58¾ Ft., WEIGHT: 59 TONS.

This was a more powerful version of the R class with a much larger boiler and a total grate area of 27 square feet. J.S. McLean records that they were built at Darlington and that the first engine, 1237, came out in November 1908 and ran from Newcastle to Edinburgh and back *as a preliminary canter. With the unexpected result that it was taken into Gateshead Works and its motion reassembled. The amusing sequel was that Gateshead sent assistance to Darlington to aid the completion of the remaining engines.* They were not as successful as the class R and the last was withdrawn in 1946.
NUMBERS: 1237-1246.

43. CLASS: S, NUMBER 760, BUILT: 1906, SHED 1920: Hull, WITHDRAWN: 1931, CLASS INTRODUCED: 1899, NUMBER IN CLASS: 40.
LNER CLASS: B13, WHEEL ARRANGEMENT: 4-6-0, DRIVING WHEEL DIAMETER: 6 Ft., LENGTH: 61 Ft., WEIGHT: 63 TONS.
The class S marked the first appearance of a 4-6-0 passenger locomotive in the British Isles. They were originally designed with a short wheelbase and this restricted the size of the cab which, on the first three engines, only had one side window. The rest of the class had the standard cab and the earlier ones were later rebuilt. For a time they handled the express passenger trains but were soon relegated to express goods and excursion traffic. The last engines were withdrawn in 1938.
NUMBERS: 726, 738-41, 743-63, 766, 768, 775, 1077 and 2001-10.

44. CLASS: S1, NUMBER: 2111, BUILT: 1900, SHED 1920: Gateshead, WITHDRAWN: 1929, CLASS INTRODUCED: 1900, NUMBER IN CLASS: 5.
LNER CLASS: B14, WHEEL ARRANGEMENT: 4-6-0, DRIVING WHEEL DIAMETER: 6¾ Ft., LENGTH: 62¼ Ft., WEIGHT: 67 TONS.
A larger-wheeled version of the class S with some improvements in design and such a tall brass safety-valve cover that its lever could only be taken through the roof of the cab. J.S. McLean records that the locomotive, pictured above, once hauled the Royal Train: *On the return of King Edward VII from his autumn visit in Scotland on October 11th 1902 the royal train was drawn by engine 2111 between Berwick and Newcastle. Mr. Wilson Worsdell, Chief Mechanical Engineer, rode on the footplate.* The last engine was withdrawn in 1931.
NUMBERS: 2111-2115.

45. CLASS: S2, NUMBER: 797, BUILT: 1912, SHED 1920: Heaton, WITHDRAWN: 1937, CLASS INTRODUCED: 1911, NUMBER IN CLASS: 20.
LNER CLASS: B15, WHEEL ARRANGEMENT: 4-6-0, DRIVING WHEEL DIAMETER: 6 Ft., LENGTH: 61 Ft., WEIGHT: 68¾ TONS.
These engines were a development of class S with wheels of the same size but a 5 ft. 6 inch diameter boiler. They were designed by Vincent Raven soon after he became Chief Mechanical Engineer, in 1910. The first seven engines used saturated steam, as was usual up to that time, but the remaining engines were fitted with superheaters which greatly increased their efficiency. The photograph shows 797, the first to be fitted with a superheater. J.S. McLean records that the first six engines were painted green but that all were repainted black with red lining.
NUMBERS: 782, 786-788, 791, 795-799, 813, 815, 817, 819-825.

46. CLASS: S2 (Stumpf), NUMBER: 825, BUILT: 1913, SHED 1920: Heaton, WITHDRAWN: 1944, CLASS INTRODUCED: 1911, NUMBER IN CLASS: 1.
LNER CLASS: B15, WHEEL ARRANGEMENT: 4-6-0, DRIVING WHEEL DIAMETER: 6 Ft., LENGTH: 61 Ft., WEIGHT: 71¾ TONS.

This was an experimental version of the S2 class which utilised the Stumpf, Uniflow, arrangement of steam distribution. The strange appearance of the engine is caused by the unusual length of the external cylinders with their piston valves mounted on top. The Stumpf design avoids the backward motion of steam in the cylinder as the long pistons uncover exhaust ports in the centre of the cylinders. J.W. Armstrong recalled that 'Stumpy' sounded more like gunfire than a locomotive and caused some consternation during the First World War! The engine was rebuilt to standard S2 in 1924 and scrapped in 1944.

47. CLASS: S3, NUMBER: 846, BUILT: 1920, SHED 1920: Gateshead, WITHDRAWN: 1963, CLASS INTRODUCED: 1919, NUMBER IN CLASS: 70.
LNER CLASS: B16, WHEEL ARRANGEMENT: 4-6-0, DRIVING WHEEL DIAMETER: 5¾ Ft., LENGTH: 62½ Ft., WEIGHT: 77¾ TONS.
A very powerful class of locomotive with three cylinders working on the simple system. The engines had smaller diameter wheels than the class S2 and put in some good work on summer excursion traffic as well as on fast goods. Thirty-five were built in North Eastern days and the same number by the LNER. Seven of the engines were rebuilt by Sir Nigel Gresley with Walschaerts valve gear, and seventeen by Edward Thompson, and many of the North Eastern-built locomotives survived into the 1960's.
NUMBERS: 840-49, 906, 908, 909, 911, 914, 915, 920-34, 936, 937, 942, 943 and LNER: 1371-85, 2363-82.

48. CLASS: T, NUMBER: 2122, BUILT: 1901, SHED 1920: Newport, WITHDRAWN: 1950, CLASS INTRODUCED: 1901, NUMBER IN CLASS: 40.
LNER CLASS: Q5, WHEEL ARRANGEMENT: 0-8-0, DRIVING WHEEL DIAMETER: 4½ Ft., LENGTH: 57½ Ft., WEIGHT: 61 TONS.
Introduced by Wilson Worsdell this was the first eight-wheels coupled locomotive class on the North Eastern. J.S. McLean records that: *These engines were designed to haul trains of 60 loaded wagons of coal. The 'Engineer' described an experiment at Tyne Dock when one of these engines hauled a train-load of 1,326 tons, whose total length was 569 yards, the longest the company had ever hauled.* The first ten engines were originally built without sand-boxes over the leading wheels.
NUMBERS: 83, 162, 410, 411, 430, 443, 444, 474, 650, 651, 715, 785, 792, 1009, 1110, 1111, 1128, 1149, 1150, 1173, 1186, 1218, 1669, 1682, 1684, 1685, 1694, 1696, 1731, 1757, 2116-25.

49. CLASS: T1, NUMBER: 669, BUILT: 1911, SHED 1920: Starbeck, WITHDRAWN: 1951, CLASS INTRODUCED: 1902, NUMBER IN CLASS: 50.
LNER CLASS: Q5, WHEEL ARRANGEMENT: 0-8-0, DRIVING WHEEL DIAMETER: 4½ Ft., LENGTH: 57½ Ft., WEIGHT: 61 TONS.
The T1 class differed only from class T in having ordinary slide valves where the latter had piston valves. Externally they were virtually the same and comparative tests showed that there was little difference in performance although fuel consumption and maintenance costs were lower for class T1. During the First World War all the engines of this class were shipped to France. They were decorated with Royal Engineers insignia and chevrons after they returned and these can just be seen on the cab side.
NUMBERS: 130, 527, 578, 642-48, 652-61, 669, 764, 767, 769-74, 781, 783, 789, 793, 794, 939, 1002, 1031, 1032, 1054, 1062, 1177, 1178, 1215, 1320, 1700, 1704, 1708, 1709, 1717, 1729.

50. CLASS: T2, NUMBER: 2259, BUILT: 1920, SHED 1920: Gateshead, WITHDRAWN: 1963, CLASS INTRODUCED: 1913, NUMBER IN CLASS: 120.
LNER CLASS: Q6, WHEEL ARRANGEMENT: 0-8-0, DRIVING WHEEL DIAMETER: 4½ Ft., LENGTH: 59½ Ft., WEIGHT: 66 TONS.
A further development of the class T, of Wilson Worsdell, introduced by Vincent Raven. This was a very successful design which was capable of developing a 1,000 horsepower pull on the drawbar at nearly 20 miles per hour. This was proved by dynamometer car trials on the Newport-Shildon branch in 1915. Fifty of the class were built for the North Eastern by Armstrong, Whitworth & Co of Newcastle in 1919 at a cost of about £8,000 per engine. All the engines lasted until the 1960's.
NUMBERS: 1247-54, 1257, 1261, 1262, 1264, 1271, 1276, 1278-80, 1283-85, 1288, 1291-94, 1311, 1335, 1361-63, 2213-2302.

51. CLASS: T3, NUMBER: 903, BUILT: 1919, SHED 1920: Blaydon, WITHDRAWN: 1962, CLASS INTRODUCED: 1919, NUMBER IN CLASS: 15.
LNER CLASS: Q7, WHEEL ARRANGEMENT: 0-8-0, DRIVING WHEEL DIAMETER: 4½ Ft., LENGTH: 61¼ Ft., WEIGHT: 71½ TONS.
This powerful three-cylinder type was intended for heavy freight or mineral traffic. Five engines were completed by the North Eastern in 1919 and ran between Newcastle and Carlisle. Ten more were built by the LNER in 1924. The engines were all superheated and were capable of hauling 1,400 tons up a gradient of 1 in 200. It was the most powerful locomotive built by the North Eastern up to that time and O.S. Nock records that one of them managed to stretch the drawbar spring, on the Dynamometer Car, to its maximum extent – a pull of 16¼ tons. Photograph: Ken Nunn/LCGB Collection.
NUMBERS: 624-26, 628-34, 901-905.

52. CLASS: U, NUMBER: 1317, BUILT: 1902, SHED 1920: Leeds, WITHDRAWN: 1959, CLASS INTRODUCED: 1902, NUMBER IN CLASS: 20.

LNER CLASS: N10, WHEEL ARRANGEMENT: 0-6-2, DRIVING WHEEL DIAMETER: 4½ Ft., LENGTH: 35½ Ft., WEIGHT: 58 TONS.

Introduced by Wilson Worsdell these were the tank engine equivalent of the P1 goods class. Problems arose when Ross 'pop' safety valves ousted the large brass safety-valve casings as J.W. Armstrong records: *Escaping steam had a habit of blowing across the cab windows so with the usual locomens ingenuity, various makeshift 'trumpets' were devised and engines would arrive at Works with such adornments as buckets with bottoms hacked out, oil drums etc., stuck round the offending item until a plain black iron trumpet was devised and fitted at the Works.*

NUMBERS: 89, 429, 1109, 1112, 1132, 1138, 1148, 1317, 1321, 1667, 1683, 1697, 1699, 1706, 1707, 1710, 1711, 1716, 1774, 1785.

53. CLASS: 66, NUMBER: 66, BUILT: 1869, SHED 1920: Darlington, WITHDRAWN: 1934, REBUILT: 1886, NUMBER IN CLASS: 1.
LNER CLASS: - , WHEEL ARRANGEMENT: 2-2-2, DRIVING WHEEL DIAMETER: 5½ Ft., LENGTH: 32¾ Ft., WEIGHT: 37 TONS.
The story of Aerolite is a remarkable one. According to J.S. McLean this locomotive is a direct descendant of a 'fine little six-wheeled single tank engine' which was painted blue and exhibited at the Great Exhibition of 1851. Other authorities say that it originates from 1869 when it *replaced* the engine built in 1851 which had been involved in 'a collision at Otterington in 1868 when attached to Mr. Fletcher's special train.' An illustrated history of the engine is given in the Journal of the Stephenson Locomotive Society, September 1951, by J.M. Fleming. The photograph shows it as a 2-2-2 but after...

54. CLASS: 66, NUMBER: 66, BUILT: 1869, SHED 1920: Darlington, WITHDRAWN: 1934, REBUILT: 1902, NUMBER IN CLASS: 1.
LNER CLASS: X1, WHEEL ARRANGEMENT: 2-2-4, DRIVING WHEEL DIAMETER: 5½ Ft., LENGTH: 32¾ Ft., WEIGHT: 44½ TONS.
...T.W. Worsdell rebuilt the engine in 1886. It was again rebuilt, this time by Wilson Worsdell in 1902, on the Worsdell-von Borries two-cylinder compound system, as a 2-2-4T. This was achieved by reversing the frames but it is doubtful if there is much of the original locomotive left. The name Aerolite was replaced by Wilson Worsdell in 1907 and the engine was used to haul the Chief Mechanical Engineer's saloon until 1933. The locomotive is now preserved in the National Railway Museum at York. This, in the author's opinion, is the North Eastern engine most worthy of restoration to working order.

55. CLASS: 190, NUMBER: 1679, BUILT: 1860, SHED 1920: York, WITHDRAWN: 1931, CLASS INTRODUCED: 1894, NUMBER IN CLASS: 2.
LNER CLASS: X3, WHEEL ARRANGEMENT: 2-2-4, DRIVING WHEEL DIAMETER: 6½ Ft., LENGTH: 32½ Ft., WEIGHT: 48¼ TONS.
The two locomotives of this class were rebuilt, in 1894, from 2-2-2 tender engines. They were larger than Aerolite, but with the same wheel arrangement, and also intended for use on Officer's Specials. An interesting note and photograph appear in the NER Magazine of 1916, page 196, which records the occasion on which 1679 became a 'Poaching Special': *Mr. C.F. Bengough, who, on July 10th, was travelling by the engineer's special from York to West Hartlepool, found on arriving at the latter place a cock pheasant tightly wedged in the front coupling of the engine.* It had evidently risen just as the engine overtook it!
NUMBERS: 190, 1679.

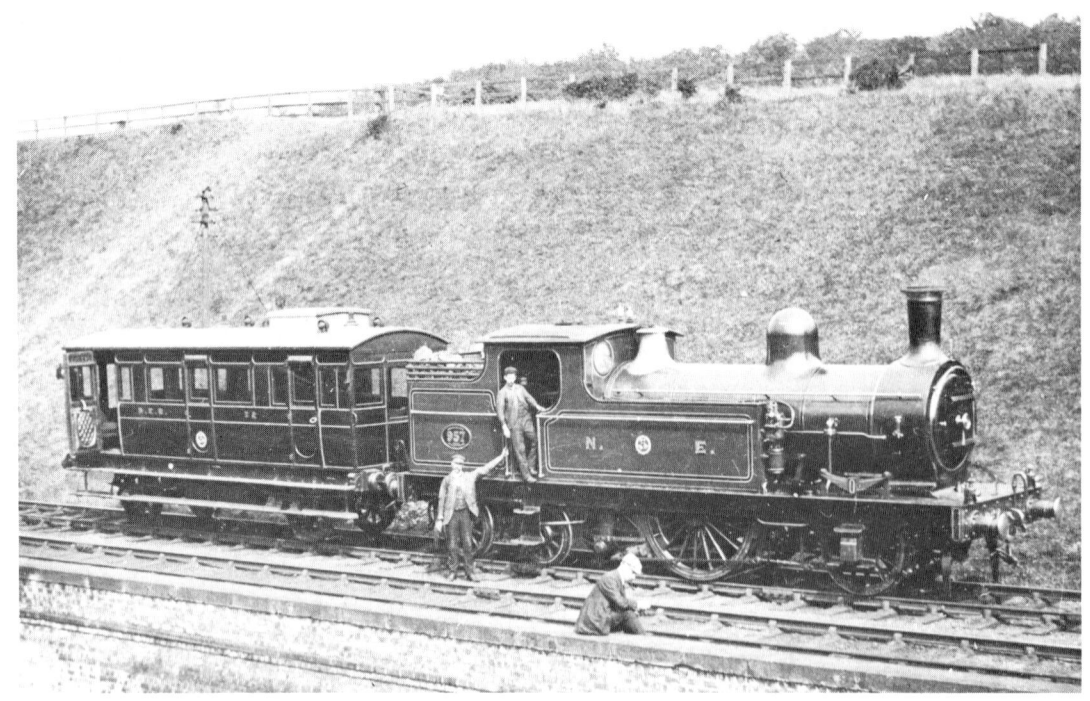

56. CLASS: 957, NUMBER: 957, BUILT: 1874, SHED 1920: Hull, WITHDRAWN: 1937, CLASS INTRODUCED: 1903, NUMBER IN CLASS: 1.
LNER CLASS: X2, WHEEL ARRANGEMENT: 2-2-4, DRIVING WHEEL DIAMETER: 6 Ft., LENGTH: 34 Ft., WEIGHT: 51 TONS.
This engine was one of a batch of 0-4-4T's originally built in 1874. It was extensively rebuilt in 1903 when a locomotive was required for pulling one of the inspection saloons. Instead of a simple rebuild 957 got the full treatment and the result was one of the finest looking tank engines on the North Eastern. In the photograph it is seen at Prospect Hill near Whitby in 1903/04 with one of the handsome four-wheeled inspection saloons. It was based at Hull for many years and could sometimes be seen on two coach passenger trains when the railcars failed in the mid-1930's.

57. CLASS: - , NUMBER: 3711, BUILT: 1908, SHED 1920: Darlington, WITHDRAWN: 1939, CLASS INTRODUCED: 1908, NUMBER IN CLASS: see below.
LNER CLASS: - , WHEEL ARRANGEMENT: - , DRIVING WHEEL DIAMETER: 3 Ft., BODY LENGTH: 17 Ft., WEIGHT: 6½ TONS.
This was the first petrol-driven inspection car on the North Eastern. It was propelled by a four-cylinder engine driving one pair of wheels through a gear box giving speeds of 15, 30 and 45 mph. Twenty gallons of petrol were carried and the car averaged 12 miles per gallon. The vehicle could be driven from either end and carried up to eight people. A new engine was fitted at York in 1929 after the car had run over 100,000 miles. Two further, larger, cars were built in 1912 — see NER Magazine 1912 (pp 140-1) and K. Hoole 'North Eastern Railway Buses, Lorries & Autocars'.

58. CLASS: V, NUMBER: 532, BUILT: 1903, SHED 1920: Heaton, WITHDRAWN: 1943, CLASS INTRODUCED: 1903, NUMBER IN CLASS: 10.
LNER CLASS: C6, WHEEL ARRANGEMENT: 4-4-2, DRIVING WHEEL DIAMETER: 6¾ Ft., LENGTH: 62¼ Ft., WEIGHT: 72¾ TONS.
The 'Atlantic' wheel arrangement was first seen on the North Eastern with the introduction of this class in 1903. It was said to stem directly from Wilson Worsdell's visit to Philadelphia in the USA. Engine 532 was the first to be built and had a low roof to the cab but as the footplate staff had difficulty in seeing over the very large boiler subsequent engines were built with raised roofs and the cab on 532 was rebuilt. The engines originally worked the East Coast expresses and the last was withdrawn in 1948.
NUMBERS: 295, 532, 649, 742, 784, 1680, 1753, 1776, 1792, 1794.

59. CLASS: V/09, NUMBER: 696, BUILT: 1910, SHED 1920: Gateshead, WITHDRAWN: 1947, CLASS INTRODUCED: 1910, NUMBER IN CLASS: 10.
LNER CLASS: C6, WHEEL ARRANGEMENT: 4-4-2, DRIVING WHEEL DIAMETER: 6¾ Ft., LENGTH: 62¼ Ft., WEIGHT: 72¾ TONS.
The V/09 class, designed in 1909 and referred to by some writers as V1, was the first to appear under the direction of Vincent Raven. It was basically the same as the class V, introduced by Wilson Worsdell, but had some alterations to the original design including strengthened mainframes. The introduction of this class also marks the changeover of the North Eastern Locomotive Headquarters from Gateshead to Darlington. The engines originally worked out of York and Leeds on express passenger services. The last one was withdrawn in 1947. Photograph: courtesy NRM, York.
NUMBERS: 696-705.

New Electric Train. N. E. R.
Commenced running march 1904.

60. The opening of electric tramways in the Newcastle area at the turn of the century seriously affected North Eastern passenger traffic returns. In an attempt to combat the competition the North Eastern electrified nearly 40 miles of track on the three-rail system with a pressure of 600 volts DC. This commenced operation in 1904, and at the same time as a similar system on the Lancashire & Yorkshire Railway, was the first outside London. The original stock was painted red and cream and had clerestory roofs but in 1918 a fire at the Heaton Carriage Sheds destroyed many of the vehicles and the replacement stock had elliptical roofs and the normal coaching livery of crimson lake. An interesting account is given in C.J. Allen's 'The North Eastern Railway'. The last of the trains ran in 1967 but the recent opening of the Tyneside Metro has revived this system.

61. In the search for economy on short branch lines the North Eastern experimented with the internal combustion engine. They also experimented with many forms of road transport but that is outside the scope of this book. Two electric autocars were built in 1903 in which the petrol engine turned a dynamo the current from which drove motors coupled to the two axles of the powered bogie. Due to technical problems with the original petrol engines the vehicles did not enter into traffic until 1904. At first one worked between Scarborough and Filey in the summer months and one at Hartlepool. They were transferred to the Selby-Cawood branch in 1908 where at least one stayed until both were withdrawn in 1930/31. The photograph shows electric autocar number 3170 as it appeared on a postcard which is postmarked 'Darlington, 1904'.

62. CLASS: BTP, NUMBER: 1431, BUILT: 1875, SHED 1920: Scarborough, WITHDRAWN: 1951, CLASS INTRODUCED: 1874, NUMBER IN CLASS: 124. LNER CLASS: G6, WHEEL ARRANGEMENT: 0-4-4, DRIVING WHEEL DIAMETER: 5 Ft., LENGTH: 33 Ft., WEIGHT: 46 TONS.

The BTP (Bogie Tank Passenger) engines were a very numerous class with a remarkable history of rebuilding and modification. Their wheel diameters, alone, varied from 5 to 5¾ feet. Although considerably 'Worsdellised' the engines still showed their early North Eastern ancestry. The photograph of 1431 is included here in conjunction with the photograph of the autocar, opposite, and also for comparison with photographs 38 and 39. Probably the most interesting books on North Eastern locomotives are the two works by J.S. McLean: 'Locomotive History of the North Eastern Railway', 1905, and 'The Locomotives of the North Eastern Railway', 1923.

63. Steam autocars were an interesting experiment carried out with some success on the North Eastern in 1905. They utilised old tank engines which were modified and coupled to special coaches with a driving compartment at one end with characteristic porthole windows. When the coach was leading the driver could control the train from the first compartment. The bogie-mounted carriage was much lighter than its equivalent train of six-wheeled coaches and the units did not have to be turned at terminal stations. Some of the units consisted of an engine and one coach, others, like the one illustrated, had a coach at each end. Many ran short-distance shuttle services but others, like the one at Whitby in 1905, ran over 86 miles every day, running out on three different lines. See G.W.J. Potter's 'A History of the Whitby & Pickering Railway'.

64. CLASS: ex901, NUMBER: 933, BUILT: 1873, SHED 1914: Barnard Castle, WITHDRAWN: 1914, ENGINE REBUILT: 1907, NUMBER IN CLASS: 1.
LNER CLASS: w/d, WHEEL ARRANGEMENT: 4-4-0, DRIVING WHEEL DIAMETER: 7 Ft., LENGTH: 53¾ Ft., WEIGHT: 53 TONS.
This was one of the famous Fletcher 2-4-0 passenger engines which was selected for special rebuilding by Wilson Worsdell in 1907. It was converted to the 4-4-0 wheel arrangement and was otherwise generally 'Worsdellised' which removed most of its earlier era characteristics although it was still coupled to a Fletcher tender! Some of the other, unrebuilt engines of the 901 class survived into LNER days, the last being withdrawn in 1925. Number 933 cannot have been a great success as it lasted only eight years after being rebuilt. See 'The 4-4-0 Classes of the North Eastern Railway' by K. Hoole.

65. CLASS: 4CC, NUMBER: 731, BUILT: 1906, SHED 1920: Gateshead, WITHDRAWN: 1933, CLASS INTRODUCED: 1906, NUMBER IN CLASS: 2.
LNER CLASS: C8, WHEEL ARRANGEMENT: 4-4-2, DRIVING WHEEL DIAMETER: 7 Ft., LENGTH: 63½ Ft., WEIGHT: 73 TONS.

The two engines of this class were built to the designs of W.M. Smith. They were four-cylinder compound engines with two high-pressure cylinders outside and two low pressure cylinders inside the frames. These were the only North Eastern locomotives to have Belpaire fireboxes and 731 was the only one built new with Walschaerts valve gear. The engines handled the express passenger traffic on the East Coast main line. Number 731 was withdrawn in 1933 and 730 in 1935. Another classic in railway literature, which deals partly with the North Eastern, is 'Locomotive & Train Working in the Latter Part of the Nineteenth Century' by E.L. Ahrons.

66. CLASS: W, NUMBER: 695, BUILT: 1908, SHED 1920: Starbeck, WITHDRAWN: 1950, CLASS INTRODUCED: 1907, NUMBER IN CLASS: 10.

LNER CLASS: -, WHEEL ARRANGEMENT: 4-6-0, DRIVING WHEEL DIAMETER: 5 Ft., LENGTH: 40¼ Ft., WEIGHT: 69¾ TONS.

The class W, nicknamed 'Whitby Willies', were powerful engines designed for the hilly branch lines on the Yorkshire coast between Scarborough and Saltburn. The first five, built in 1907, had extended smokeboxes, and all had variable blastpipes which reduced the amount of sparks and cinders flying out of the chimney when the engines were working hard uphill. All the engines had a bunker which held only 2¼ tons of coal, except for 695 which carried 3 tons. Some bridges on the coast route had to be strengthened to carry the engines so most were at first allocated to the Starbeck-Leeds area.

NUMBERS: 686-695.

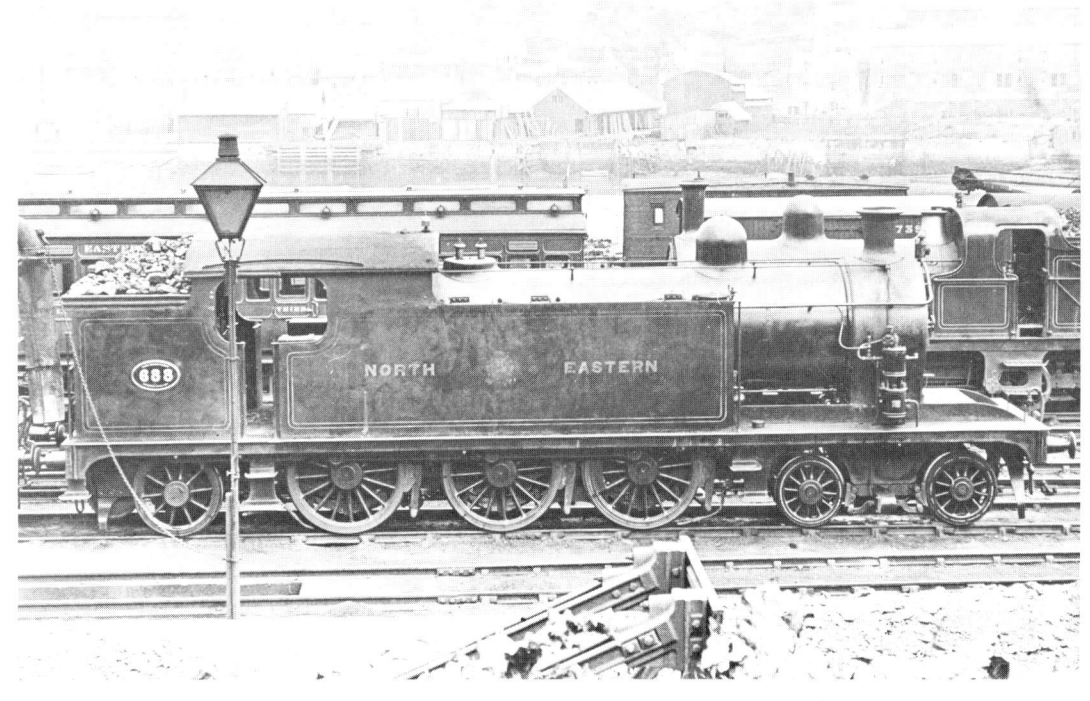

67. CLASS: W, NUMBER: 688, BUILT: 1907, SHED 1920: Leeds, WITHDRAWN: 1948, REBUILT: 1914, NUMBER IN CLASS: 10.
LNER CLASS: A6, WHEEL ARRANGEMENT: 4-6-2, DRIVING WHEEL DIAMETER: 5 Ft., LENGTH: 43¾ Ft., WEIGHT: 78 TONS.

Considering that these engines were designed for strenuous work, implying the consumption of a higher than average amount of coal, it is surprising that Wilson Worsdell provided them with such small bunkers. Sir Vincent Raven soon altered them, from 1914 to 1917, to carry 4 tons of coal and provided a pair of trailing wheels to carry the extra length and weight. They were also unusual, for Wilson Worsdell locomotives, in having inside cylinders and one is tempted to wonder if they were designed and built whilst he was on holiday! The last engine was withdrawn from Hull in 1953. The photograph shows 688 in Whitby yard.

68. CLASS: X, NUMBER: 1352, BUILT: 1909, SHED 1920: Hull, WITHDRAWN: 1959, CLASS INTRODUCED: 1909, NUMBER IN CLASS: 10.
LNER CLASS: T1, WHEEL ARRANGEMENT: 4-8-0, DRIVING WHEEL DIAMETER: 4½ Ft., LENGTH: 42 Ft., WEIGHT: 85 TONS.
A very powerful three-cylinder design for heavy shunting and marshalling in 'hump' yards at Hull, on Teesside and other docks. It is recorded that, on test, one of these engines hauled a train weighing 1,200 tons, from Hull to Bridlington. These were the last new engines built at Gateshead Works and the three cylinders and steam chests were cast in one block. Five more engines were built by the LNER in 1925, with Ross 'pop' safety valves. The last North Eastern engine was withdrawn in 1959.
NUMBERS: 1350-1359, 1656-1660.

69. CLASS: Y, NUMBER: 1113, BUILT: 1910, SHED 1920: York, WITHDRAWN: 1954, CLASS INTRODUCED: 1910, NUMBER IN CLASS: 20.
LNER CLASS: A7, WHEEL ARRANGEMENT: 4-6-2, DRIVING WHEEL DIAMETER: 4½ Ft., LENGTH: 44 Ft., WEIGHT: 87 TONS.
These tank engines were designed for heavy mineral traffic and had a relatively flexible wheelbase to allow them to work into short-radius colliery and dock sidings. The three cylinder arrangement was again used but the boilers were larger and the cylinders smaller than the class X. They were designed to pull a gross load of 1,000 tons on the level at 20 miles per hour. The last engine was withdrawn in 1957.
NUMBERS: 1113, 1114, 1126, 1129, 1136, 1170, 1174-76, 1179-83, 1185, 1190-93, 1195.

70. CLASS: Z, NUMBER: 716, BUILT: 1911, SHED 1920: Gateshead, WITHDRAWN: 1948, CLASS INTRODUCED: 1911, NUMBER IN CLASS: 10.
LNER CLASS: C7, WHEEL ARRANGEMENT: 4-4-2, DRIVING WHEEL DIAMETER: 6¾ Ft., LENGTH: 63¼ Ft., WEIGHT: 79 TONS.
In 1911 ten three-cylinder saturated engines of class Z were built at the same time as ten superheated engines of class Z1. The superheated engines performed better and the ten saturated engines were equipped with superheaters within four years and all became class Z. The twenty engines were all built to the designs of Vincent Raven by the North British Locomotive Company of Glasgow at a cost of almost £4,500 each. These original engines all had large safety-valve casings whilst subsequent engines, and rebuilds, were fitted with Ross 'pop' safety valves. The last engine was withdrawn in 1948. Photograph: Ken Nunn/LCGB collection.
NUMBERS: (class Z) 706, 709, 710, 714, 716-21. (class Z1) 722, 727-29, 732-37.

71. CLASS: Z1, NUMBER: 735, BUILT: 1911, SHED 1920: Gateshead, WITHDRAWN: 1945, CLASS INTRODUCED: 1911, NUMBER IN CLASS: 40.
LNER CLASS: C7, WHEEL ARRANGEMENT: 4-4-2, DRIVING WHEEL DIAMETER: 6¾ Ft., LENGTH: 63¼ Ft., WEIGHT: 79 TONS.
Following the original ten superheated engines of Z1 in 1911, further batches were built at Darlington from 1914 to 1918, with small modifications in boiler size and other details. The photograph shows one of the original batch. A new and improved design of tender was provided by Vincent Raven. The class as a whole was successful and versatile – J.S. McLean recorded that 732 hauled the 'Flying Scotsman' weighing 545 tons, whilst C.J. Allen recorded 2195 at nearly 80 miles an hour on the level. 2212 was built with three sets of Uniflow cylinders on a modified form of Stumpf system (see caption 46).
NUMBERS: 2163-72, 2193-2212.

72. CLASS: - , NUMBER: 2, BUILT: 1905, SHED 1920: Heaton, WITHDRAWN: 1964, CLASS INTRODUCED: 1905, NUMBER IN CLASS: 2.
LNER CLASS: ES1, WHEEL ARRANGEMENT: 0-4-4-0, DRIVING WHEEL DIAMETER: 3 Ft., LENGTH: 38 Ft., WEIGHT: 56 TONS.
The two electric locomotives of this class were designed for use on the short but steep Quayside branch of the North Eastern at Newcastle-upon-Tyne. The line was nearly semicircular and less than a mile in length, much of it running through tunnels and cuttings on gradients as steep as 1 in 27. The locomotives were fitted to operate both with the overhead line or third rail on a 600 volt supply. They were capable of starting a 150 ton train on the steepest gradient. Originally they were fitted with a 'bow' collector at one end but this was later replaced by a pantograph mounted on the cab roof. One engine is preserved in the National Railway Museum.
NUMBERS: 1 & 2.

73. CLASS: - , NUMBER: 3, BUILT: 1914, SHED 1920: Shildon, WITHDRAWN: 1950, CLASS INTRODUCED: 1915, NUMBER IN CLASS: 10.
LNER CLASS: EB1, WHEEL ARRANGEMENT: 0-4-4-0, DRIVING WHEEL DIAMETER: 4 Ft., LENGTH: 39¼ Ft., WEIGHT: 74½ TONS.
It was the boast of the North Eastern Railway (NER Magazine June 1916, pp 127-129) that the Shildon-Newport section of their line was the first heavy mineral railway in the world to be worked by electric locomotives. The current in the overhead wires was supplied at 1,500 volts and the engines were designed by Vincent Raven and built at Darlington Works. Each locomotive was equipped with four electric motors, of 275 horsepower, and was capable of hauling a 1,400 ton train at 25 mph. The engines continued in use on the line until 1935 when they went into storage.
NUMBERS: 3-12.

74. CLASS: - , NUMBER: 13, BUILT: 1922, SHED: Darlington, WITHDRAWN: 1950, CLASS INTRODUCED: 1922, NUMBER IN CLASS: 1.

LNER CLASS: EE1, WHEEL ARRANGEMENT: 4-6-4, DRIVING WHEEL DIAMETER: 6½ Ft., LENGTH: 38 Ft., WEIGHT: 110 TONS.

This impressive electric locomotive was, unfortunately, way ahead of its time. It was built under the far-sighted direction of Sir Vincent Raven for express passenger traffic on the East Coast main line. The planned electrification of this line, by the North Eastern, did not materialise as, after the amalgamation of 1923, the LNER did not take up the scheme. The engine ran on a few trial trips on the Shildon-Newport line but spent most of its life stored at Darlington and was sold for scrap in 1950. Now, more than sixty years after the engine was built, the British Railways Board have announced their intention to electrify the East Coast Main Line! Photograph: courtesy NRM, York.

75. CLASS: 4-6-2, NUMBER: 2400, BUILT: 1922, WITHDRAWN: 1936, CLASS INTRODUCED: 1922, NUMBER IN CLASS: 5.

LNER CLASS: A2, WHEEL ARRANGEMENT: 4-6-2, DRIVING WHEEL DIAMETER: 6¾ Ft., LENGTH: 72½ Ft., WEIGHT: 97 TONS.

The mere sight of this monster 'Pacific' locomotive was enough to give a fireman nightmares! They were designed by Sir Vincent Raven and were the last new steam locomotive class to be built by the North Eastern. Only two were completed before the Grouping in 1923 and were still 'on trail' until April of that year. Three more were built by the LNER in 1924 but comparisons with the Great Northern Railway Pacifics were unfavourable and all were withdrawn by the end of 1937. All the engines were given the names of North Eastern cities and 2400 was named 'City of Newcastle' in 1924.

NUMBERS: 2400-2404.

76. Finally, a last look back to the halcyon days of steam — Whitby shed and yard in about 1920. Counting the domes there are ten engines on this card including two class A, in the foreground, and two class W, on the left. Immediately in front of the signal is a BTP Autocar set. Admirers of North Eastern locomotives are fascinated by the story told by E.L. Ahrons in his book 'Locomotive & Train Working in the Latter Part of the Nineteenth Century' in which he describes a 'ghost' engine shed — 'Somewhere on the North East Coast'. It contained about thirty old engines and was situated in a bleak and lonely spot within sound of the sea. He suggested that it might still exist, but a more tangible locomotive eldorado can be found within the National Railway Museum at York!